LEARN SEO IN 2019

Written by

SAJIB HASAN

Learn more from our website

www.creativezib.com

Table of Contents

Chapter 1

SEO Basics (Page-3)

Chapter 2

Search Engines (Page-9)

Chapter 3

On-Page SEO (Page-19)

Chapter 4

Content & SEO (Page-31)

Chapter 5

Keyword Research (Page-40)

Chapter 6

Link Building (Page-48)

Chapter 7

UX and SEO (Page-56)

Chapter 8

SEO Resources (Page-64)

Chapter 1

SEO Basics

Before we dive into specific techniques and aspects of how to learn SEO in 2019, let's start with the first chapter which deals with the basic terms. Welcome to the ultimate SEO guide for beginners!

Are you new to SEO? Do you wonder how it works and what matters most in 2019? You're at the right place!

What is SEO?

Search engine optimization (SEO) is a method of improving positions in natural (non-paid) search consequences in search engines. The higher the internet site is, the greater humans see it.

The records of search engine marketing dates again to the 90s when the search engines emerged for the first time. Nowadays, it is an necessary advertising and marketing approach and a developing industry.

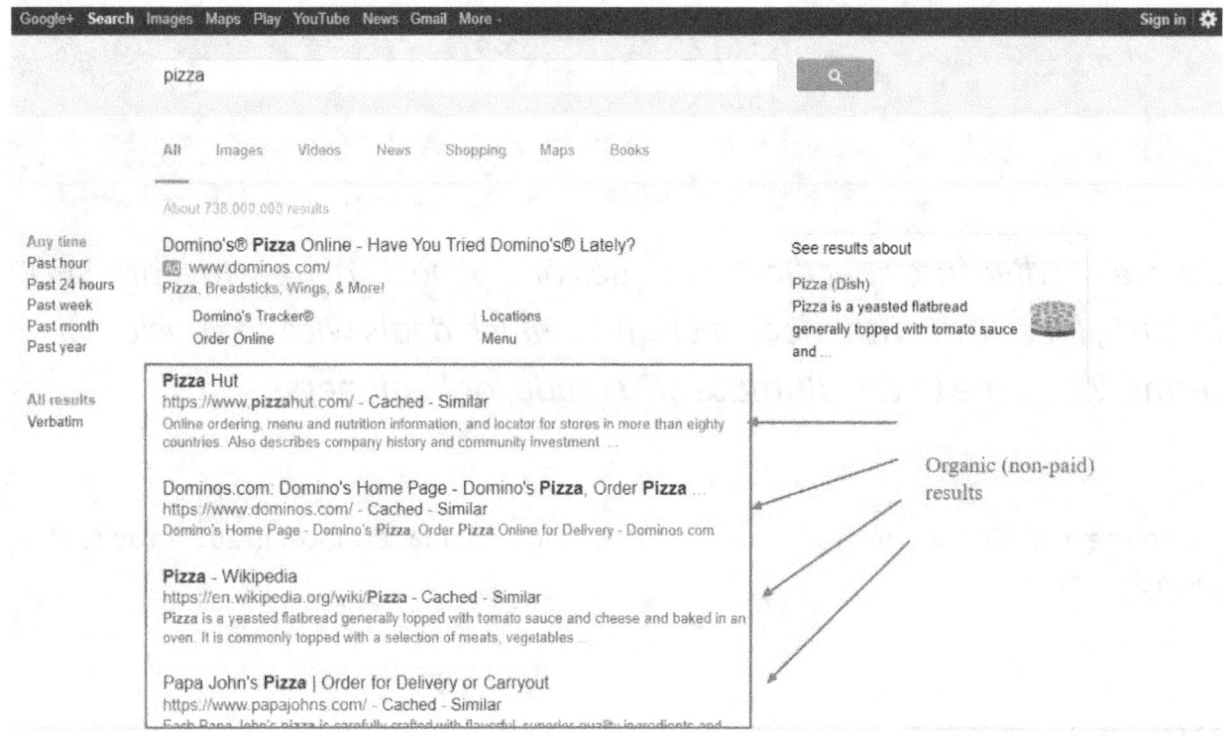

Organic search results for keyword: "pizza"

If you choose to research SEO, you must understand that it requires a lot of creative, technical and analytical work. There are many techniques with distinct goals, however, the foremost point will continue to be the equal – to be among the highest results in organic searches.

Simply said, SEO is about running the right website for the right people.

It isn't only about a ideal shape or technical heritage of the website. Your web has to be crammed with great well-optimized content material tailored to the wishes of your audience. Make sure your content is good sufficient to be linked by other websites.

SEO is a complex self-discipline that desires your full attention. And the first-class part? It evolves all the time so your website should continually maintain up with the changes.

Why should I learn SEO?

Search engines such as Google, Bing, Yahoo! and others index websites to create an order based on a number ranking algorithms. Can we become aware of these algorithms? Yes and no.

Google makes use of extra than 200 ranking factors. We be aware of many of them: content, backlinks, or technical things such as web page speed, but there are many kept as a secret.

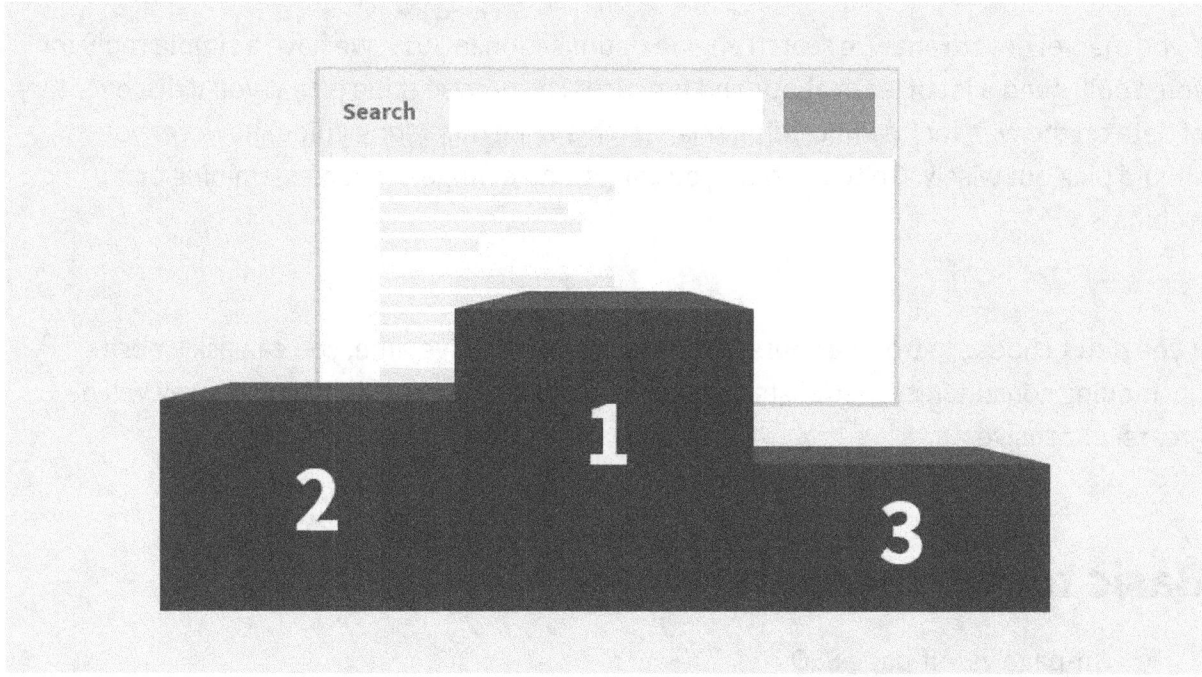

Search engines are used by the internet users when they are looking for something. And you choose to be that "something", don't you? It doesn't depend whether you promote a product, service, write a blog, or whatever else, search engine optimization is a must.

Your Website desires to be listed by means of search engines. Otherwise, you're lost.

Wise search engine marketing activities enhance your rankings in the search engine results page (SERP). Higher rankings imply higher traffic. If the traffic is engaged, it will convey conversions.

To sum it up, if you design to be successful with your website, you'll want to do SEO.

How can I learn SEO? Do I need someone's help?

Even the fundamental adjustments can make a big difference in how search engines see your website. In this final SEO guide for beginners, we'll cowl all the crucial topics and web optimization basics. You'll attain ample knowledge to proceed with search engine marketing on your own.

If you marvel how to analyze search engine optimisation in 2019, we have a simple reply for you: You'll need a lot of learn about and practice. The precise thing is that you'll discover tons of data on the web for free (including this search engine marketing guide) however you should pick out wisely. On top of that, you can attend a range of courses, training or webinars.

If you don't choose to trouble yourself that tons or don't have time, you can ask website positioning consultants, specialists or agencies for help. Keep in idea that this way will be greater expensive.

Basic terms vocabulary:

- On-page vs. off-page SEO
- White hat vs. black hat vs. grey hat SEO

On-page vs. off-page SEO

Doing On-page (on-site) SEO potential optimizing your internet site to have an effect on search engines results. It's the entirety you can do on the internet site – from content material through technical aspects to design:

- meta tags
- headings
- URL structure
- images optimization
- content
- website size and speed

Off-page (off-site) website positioning covers all things to do you do to improve the area authority via getting one way links from other websites. There are many ways to get them:

- exchanging backlinks
- guest blogging
- buying backlinks
- social media efforts
- cooperation with influencers
- writing valuable content, so people would love to link to your website

White hat vs. black hat vs. grey hat SEO

Are you a bad guy? Black hats vs. white hats have their starting place in Western movies. It's like horrific guys vs. desirable guys. But don't take these phrases too seriously. Opinions on each search engine optimization tactics have a tendency to differ.

Black hat SEO is a set of unethical practices to improve rankings of a website in the search engine consequences page. They are designed to have an effect on search engines whilst no longer taking human component into consideration.

Black hat search engine marketing can get you to the pinnacle of SERP in a brief time, however, search engines will most in all likelihood penalize and ban the internet site quicker or later.

White hat SEO is a set of ethical techniques sticking to the guidelines and rules. The basic parts of white hat SEO are:

- quality and relevant content
- link building

White hat search engine marketing is a long-term strategy oriented to client experience. Being a precise guy in the world of search engine marketing must be the desirable direction.

There's also a term called the Grey hat SEO, a practice when you may hazard less when in contrast to the Black hat techniques. Grey hat strategies aren't certainly described with the aid of Google so you can obtain heaps of internet site customers while no longer being penalized or misplaced all your rankings a day after.

Generally speaking, you don't choose to make Google your enemy.

Chapter 2

Search engines

In the 2nd chapter of our search engine marketing guide, you will examine how search engines work, how humans use them and what type of search queries they submit. Take a seem to be at the technical background behind Google.

Let's take a closer seem to be at the search engines and what are the most common ranking factors you must focus on.

How search engines work!

Search engines consist of three main ingredients:

- Crawling
- Indexing
- Picking the results

The process looks like this:

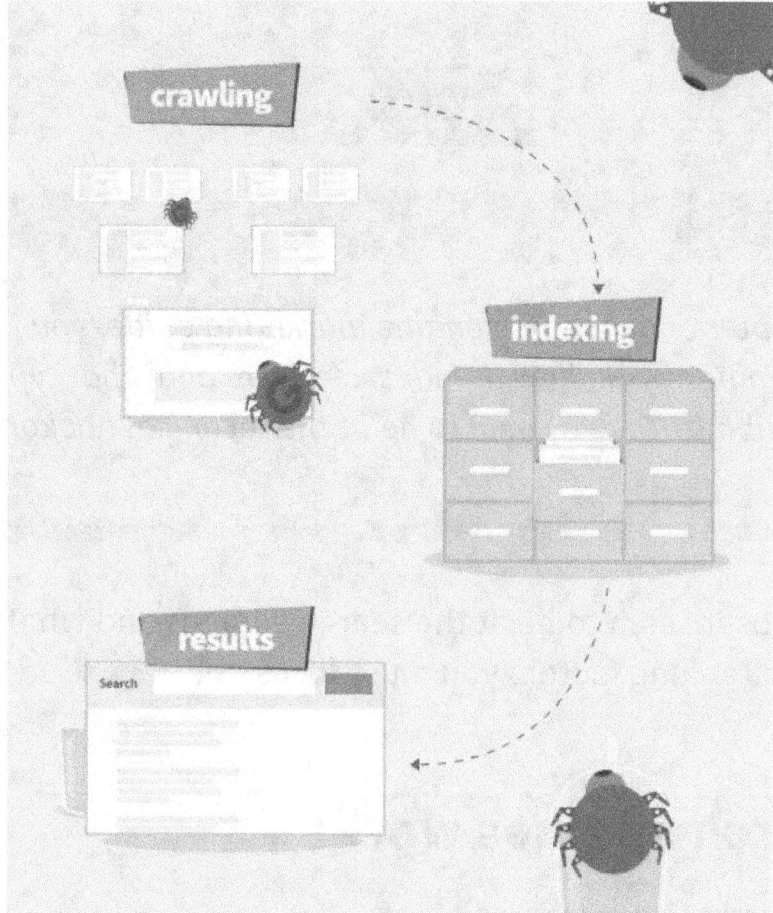

Crawling

Crawling or spidering means scanning the website, its sections, content, keywords, headings, links, photos by way of heaps of small bots. Any data that can be discovered on the internet site is crawled.

Crawlers realize any hypertext hyperlinks on a internet site pointing to different websites. Then they parse those pages for new hyperlinks over and over again. Bots crawl the total web usually to replace the data.

Indexing

Once the internet site is crawled, the indexing takes place. Imagine the index as a gigantic catalog or a library full of web sites from all over the world. It normally takes some time for a internet site to be indexed. From our experience, it's from 1 to 10 days.

Quick tip:

You can take a look at what pages have already been listed with the aid of the usage of this search operator: site:domain.com

Furthermore, every time it's changed, our proper friend crawler scans it again. Keep in mind that until updates on the internet site are indexed, they won't be visible in search engines.

Picking the results

Results are indispensable for each developers and users. Once the net user submits a search query, the search engine digs into the index and pulls out matching results. It's a manner of checking the query towards billions of websites primarily based on a variety of algorithms.

Companies strolling search engines (Google, Microsoft, Yahoo!) hold the specific calculations of their algorithms in secret. Nonetheless, many ranking factors are well-known.

Ranking factors

Most of these elements are proven, but some are simply speculations or even myths. On pinnacle of that, some are more important than others. Brian Dean from Backlinko made a fine listing of Google ranking factors.

You don't have to recognize all of them by means of heart to research SEO, however it is top to have at least a simple overview.

One of the most vital factors, the oneway link profile is primarily based on the number and high-quality of back-links leading to a website. It's a very simplified view on Google approximation of the website's authority. Each one way link is an analogy of an educational citation.

These are some of the most necessary ranking elements (in no precise order):

- Strength and relevancy of backlinks (incoming links to your website)
- Content relevancy and quality
- Engagement metrics such as CTR (click-through rate)
- Website size and loading speed
- Keyword density, keyword usage in headings, meta tags, URL – these aren't as important today as they used to be from the technical point of view
- Grammar and spelling
- Website structure
- Mobile optimization
- Overall domain authority
- Social signals
- Internal linking
- Website usability

Ranking factors can be divided into on-page SEO factors (including technical SEO) and link building or off-page SEO factors.

Quick tip:

Check domain authority of any website to find its strengths and weaknesses.

They are extremely important and deserve more attention. We focus on them in Chapter 3 and Chapter 6.

How people use search engines

The fundamental point of search engine optimization is to be friendly both to users and search engines. If you make investments all your money and time into perfect technical SEO, it's fine. But if the person interaction is poor, your positions can suffer. And that's how you start losing money. The user's factor of view is a wide variety one priority.

The photograph below represents one of the frequent person journeys in Google search:

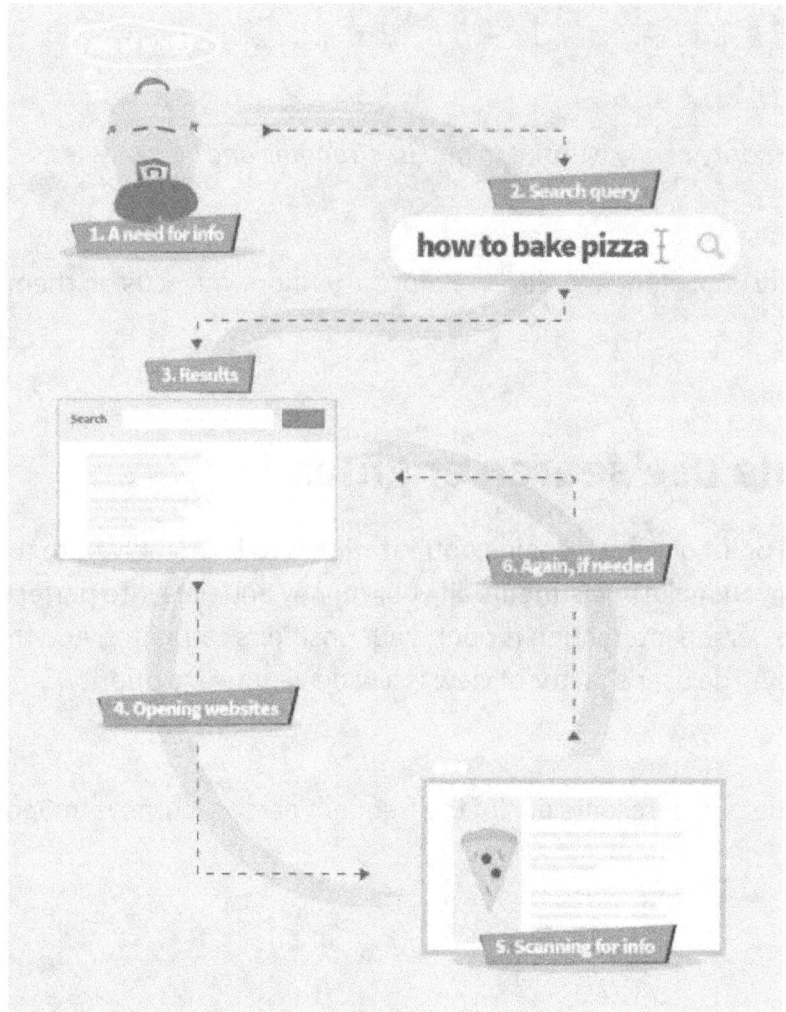

The interactions with search engines have evolved over the years. However, the principle remains the same:

- A need for a solution, information, or an answer
- Typing the need in form of a query (keyword) into the search engine
- Going through the first results
- Clicking on one or more results (websites)
- Scanning websites for the answer
- Going through more results on the 1st SERP and/or changing the search query, if the answer isn't found.

Search engines market share

In the charts below, you can see which search engines people use the most. The data is from Netmarketshare's reports.

Desktop search engine market share (Jun 2017 - May 2018)

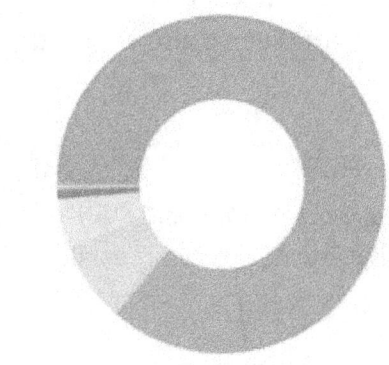

Google (72.54%) ■ Baidu (13.48%) ■ Bing (7.65%) ■ Yahoo! (4.74%) ■ Yandex (0.86%) ■ Ask (0.30%)
■ DuckDuckGo (0.22%) ■ Naver (0.12%) ■ AOL (0.06%) ■ Dogpile (0.04%)

Mobile search engine market share (Jun 2017 - May 2018)

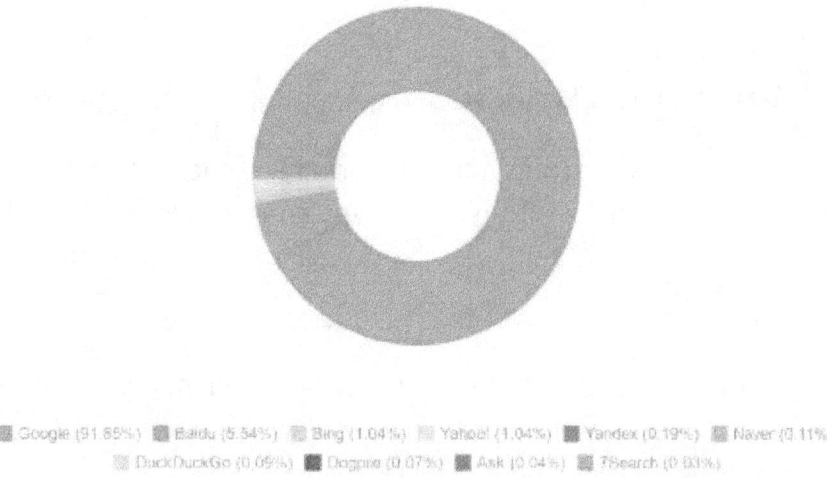

Google (91.85%) ■ Baidu (5.54%) ■ Bing (1.04%) ■ Yahoo! (1.04%) ■ Yandex (0.19%) ■ Naver (0.11%)
■ DuckDuckGo (0.09%) ■ Dogpile (0.07%) ■ Ask (0.04%) ■ 7Search (0.03%)

Find out more about the most popular search engines and their history.

How do we classify search queries?

There are three generally accepted types of search queries:

- Navigational search queries
- Informational search queries
- Transactional search queries

www.creativezib.com

Navigational search queries

They characterize an intent to search for a particular brand or website. People tend to kind "youtube" or "google" into search engines as an alternative than the use of browser's records or bookmarks.

Based on our case learn about the place we analyzed 300+ millions of keywords, YouTube, Facebook and Google attain the highest search volumes alongside with other navigational search queries.

Informational search queries

These are submitted when users are looking out for information. They aren't searching for a precise website, but for an reply or training on how to do something. For example, "How to bake pizza".

Transactional search queries

This kind is an intention to make a transaction. It normally comes with a product name (Nike Airmax) or class (sneakers). Additionally, it can be written with "Where to purchase …", "… price" or in a similar manner.

There are many weblog posts on how to target a specific search query. However, it's no longer that convenient because of the increasing popularity of voice assistants such as Siri, Google Now or Alexa.

Informational search queries can shortly seriously change to transactional by opening a new app or giving an alternative to make a purchase.

1st area vs. 1st page

Being on the first web page of the natural search outcomes is good, scoring the pinnacle three is extremely good however there's solely one winner, right? Or, is it? It's a count of perspective.

Websites all over the world are updated on a daily, weekly or monthly basis. The internet grows every single day. When new websites and modifications are listed via a search engine, the natural results might also change.

Another very vital factor is Google algorithm which adjustments all the time. Minor tweaks may additionally not reason something at all, however a main algorithm update can be an earthquake.

What we're attempting to say is that even if you're the winner, your positions can (and probable will) be replaced by opponents the other day, and vice versa.

In the chart below, you can see the importance of the easiest rankings in Google relying on their natural click-through fee (CTR) distribution for April 2018 (based on Advanced Web Ranking).

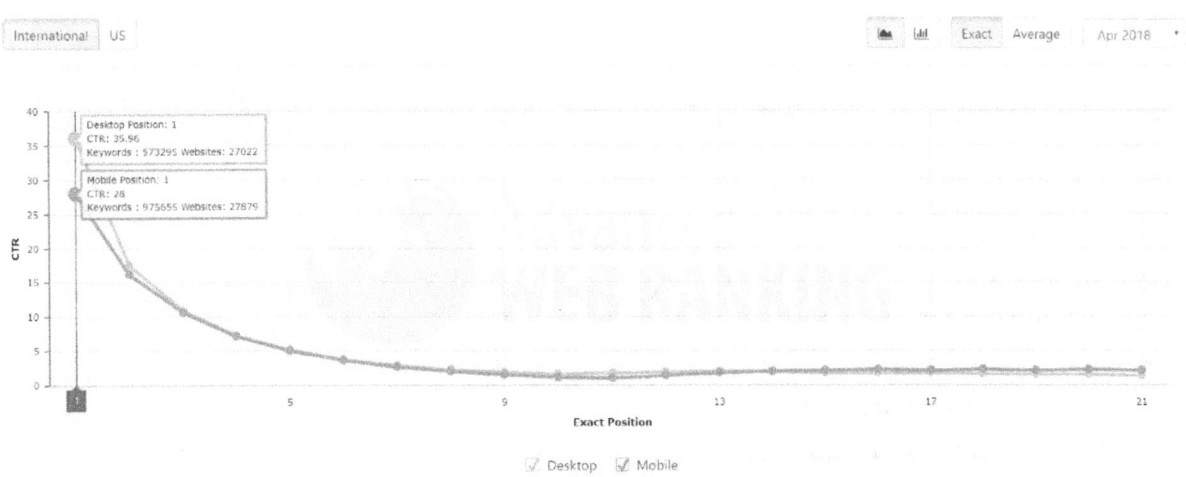

Of course, ranking first is very important, however these days, you have to take into consideration also the so-called "zero position".

Let's take a look at the outcomes for "How to bake potatoes" search query. The first result is a Google featured snippet with all the information, so you don't need to take a look at different results anymore.

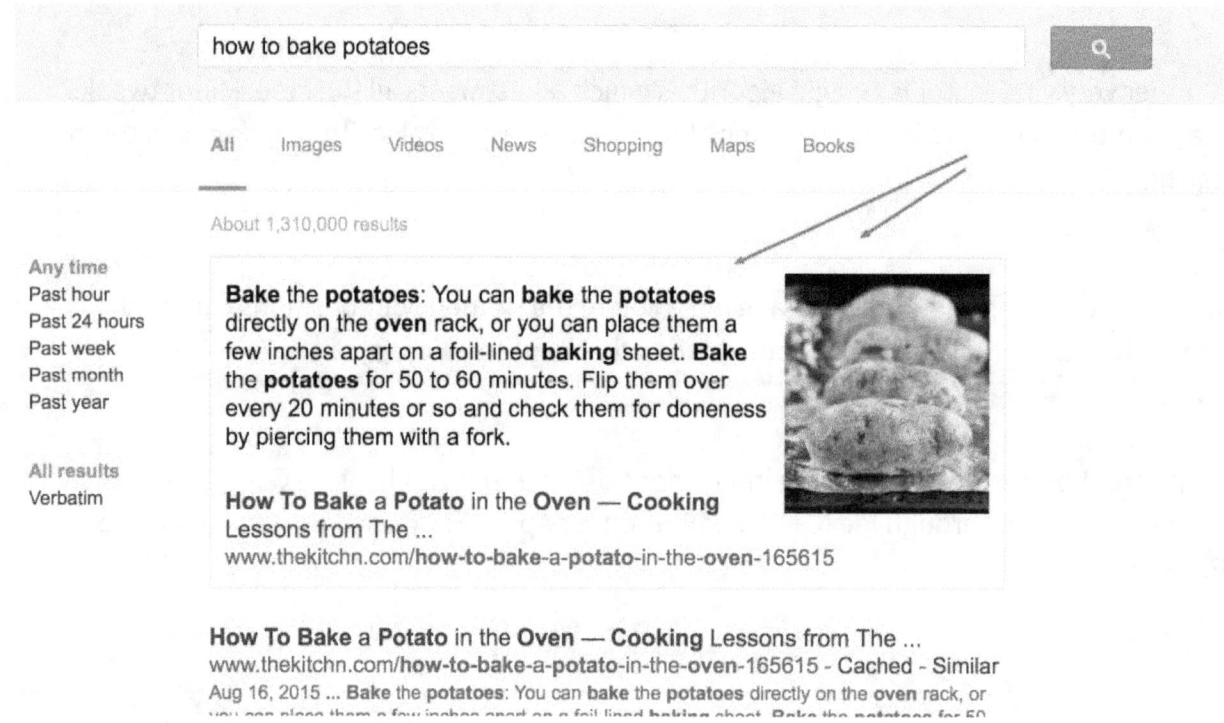

The vocabulary of the 2nd chapter:

- crawling
- indexing
- picking results
- ranking factors
- navigational search queries
- informational search queries
- transactional search queries
- featured snippets

Chapter 3

On-page and technical SEO

On-page search engine optimization consists of pretty a lengthy checklist of tasks to do. A few years ago, it was once basically about meta tags, over-optimized content and headlines. If you desire to study website positioning in 2019, you'll have a lot extra fun. Let's dive in.

Search engine algorithms have come a long way and grow to be extra sophisticated from the times when content, titles and descriptions stuffed with key phrases were adequate to gain excessive rankings in search results.

If you choose to examine search engine marketing in 2019, you can't get stuck in the past. One of the biggest enhancements of search engines is that they consider the human factor, in other words, the user engagement.

Don't forget that SEO is about targeting real people, not only search engines.

Meta tags

Meta tags are a part of the HTML code. They describe website's content. The most essential are meta titles and meta descriptions.

For distinct information, make positive to study the definitions and superior optimization recommendations in our SEOpedia each for title tags and meta descriptions.

HTML code snippet with meta tags

They aren't as essential as they used to be from the technical factor of view.

Meta titles and descriptions packed with keywords don't at once influence your rankings BUT they remain a robust psychological component affecting the CTR and universal person engagement.

So don't get stressed by way of blogs pronouncing that title tags and meta descriptions aren't necessary at all. On the different hand, hold in idea that Google modifications titles and descriptions to higher point out their relevance to the search question in case your text doesn't match enough.

There are many equipment and plugins (e.g. Yoast for WordPress in the screenshot below) analyzing your content material in phrases of focal point key-word utilization in meta title, meta description, headings, universal keyword density, alt photograph attributes and others.

They provide a lot of hints, yet can hardly ever comply with Google's algorithm focusing on person experience. Make certain that you reflect onconsideration on each technical factors of view and consumer experience when growing meta tags.

Setting title tag and meta description in the Yoast SEO plugin for WordPress

On-page SEO checklist

Let's begin with the matters you can do in WordPress or in any other content material administration system.

1. Find out what people are looking out before you begin writing

Do you design to write about a subject that human beings search for? Are you sure your factor of view is different, unique? Can you provide new brought fee to the topic? Is your timing right?

These are the major questions before you start writing and it doesn't matter whether it's a blog or a product touchdown page. There are many ways how to find out. Keyword lookup is an inevitable phase of it. We speak about how to do key-word lookup in the fifth chapter.

2. Title tags and headlines

Create an attractive title tag, meta description and headlines. Keep in thinking what we noted earlier. Your primary focal point keyword have to nonetheless be there, so customers know what is your internet site about. Use the meta description as a notable chance for call-to-action (CTA) emphasis.

Persuade each customers and search engines that your website is the one to be clicked on.

Once again, think of the consumer engagement, so don't overact through using low priced or too tacky words. Look at your competitors, analyze what works for them and construct your personal strategy.

Quick tips:

- Google will show the titles if they're up to 70 characters and meta descriptions up to 155 characters (updated on May 2018)
- Make sure to use correct <h1>, <h2>, <h3>, ... structure for good readability and structure
- Check the search results preview in tools such as SEOSiteCheckup or WordPress plugins

3. Use SEO-friendly URLs

Avoid using auto-generated URLs with figures and characters:

www.example.com/2017/post318e7a349f6

Use URLs corresponding to your content and its title:

www.example.com/how-to-bake-pizza

If you use WordPress, you can set permalinks in common settings.

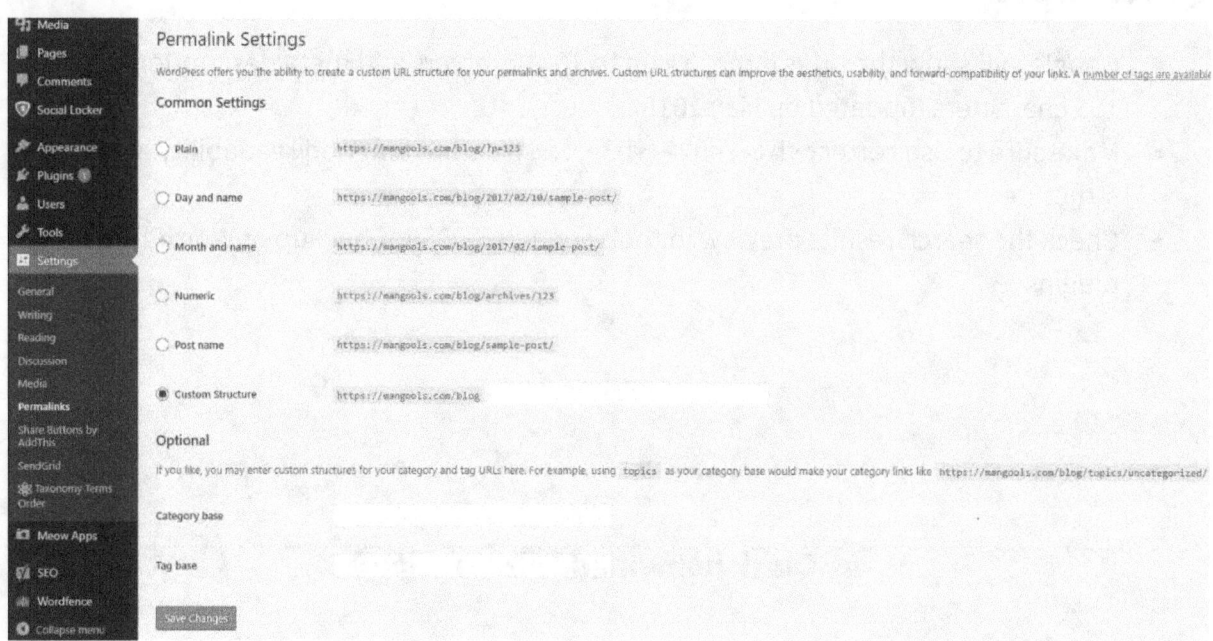

How to set up permalinks in WordPress

Some website positioning experts and bloggers say that brief URLs ranks higher in Google. We assume it's a matter of the consumer experience. Of course, this doesn't mean a 20-word URL is alright.

4. Multimedia

Do you prefer to have interaction your visitors? Use images, infographics, charts and videos. They lead to lower soar fees and a higher engagement. Some things have to be written in the desirable old fashioned way however multimedia are a must.

Video streaming has been one of the most up to date advertising and marketing trends over the last couple of years. Furthermore, they motivate people to like, share or remark your content.

Quick tips:

- Optimize images by using relevant file names (how-to-bake-pizza.jpg), alt image attributes and file size
- Embed interactive multimedia such as videos or charts
- Don't forget to include transcripts so you don't lose important keywords (the crawlers can't "read" the video)

5. Outbound and internal links

Using outbound hyperlinks gives a relevancy sign of your topic to Google. Make positive to hyperlink to applicable and authoritative sources.

Internal hyperlinks are a ideal way to promote your other articles or internet site sections. It makes less difficult to go to them and leads to greater engagement. Internal linking also helps the bots to apprehend the internet site structure.

Quick tips:

- Outbound links may not directly improve your rankings, yet it is highly advisable to use them
- Use up to 2-3 internal links, depending on the content length
- Search engine crawlers scan these links, so don't try to cheat and watch out for broken links

6. Let human beings engage

Great content material shares itself. Yes, possibly in the past. People are lazy these days so the share buttons have to constantly be on your website. According to BuzzSumo, social sharing dropped via 1/2 because 2015.

Besides Facebook, Twitter or Linkedin, reflect onconsideration on adding particular and topic-related social networks, such as Reddit or Pinterest.

Technical on-page SEO checklist

We can classify technical web optimization as a section of on-page SEO that offers with more technical stuff. It normally needs development abilities or a web developer.

1. Search console

Google Search Console (Webmaster tools) is one of the website positioning basics. It helps you to screen and keep your site's presence in Google search results.

That includes content submission for crawling, selecting what you desire to be indexed and what not, website online errors, structured information or code errors.

The Search Console will assist you to analyze your keyword rankings, CTRs, viable Google penalties and many different beneficial data for technical SEO.

Quick tips:

- Every property (website) desires to be validated to use Search console features
- Connect Search Console with Google Analytics

2. Website speed

Website pace is one of the rating factors. Slow internet site reasons that people will leave. It's acknowledged that 50% of web users count on a website online to load in 2 seconds or less. If it doesn't load in three seconds, they will leave.

Quick tips:

- Test the velocity in PageSpeed Insights
- Optimize images, enable GZIP compression, HTML compression, JS and CSS minification and attempt to limit server response time
- Quality net hosting performs a huge position in the internet site speed, so make positive to pick a honest provider

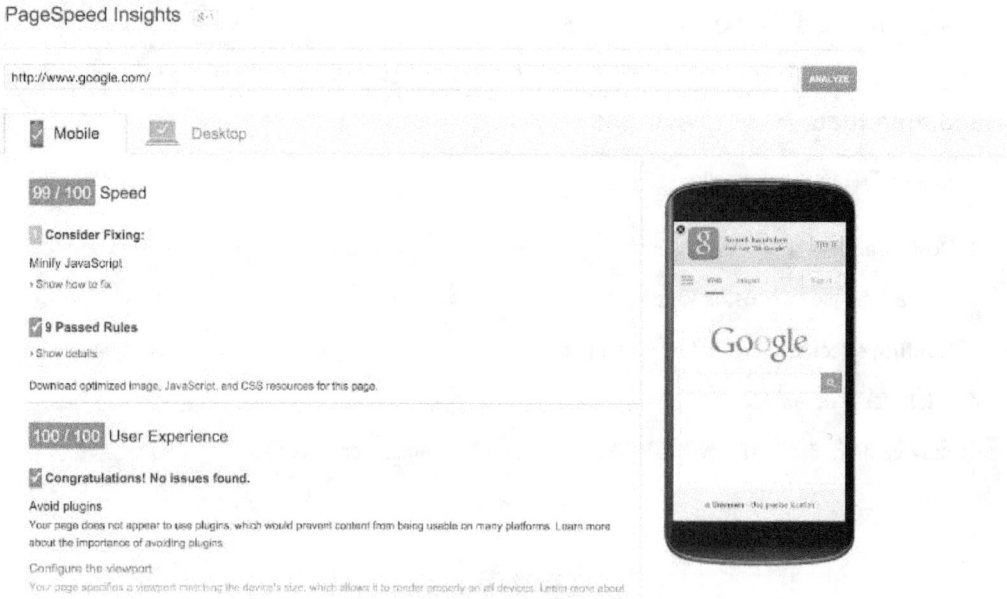

Testing site speed of your website

3. Mobile optimization

Mobile optimization is a must. The world is shifting from computing device to mobile. In fact, running a internet site that is no longer mobile-optimized will negatively have an impact on your rankings.

Google started out rolling out the mobile-first indexing in March 2018. Mobile-first indexing ability that Google will use the cellular model of your internet site for indexing and ranking.

You can additionally consider the AMP (Accelerated Mobile Pages). AMP is an HTML code extended with custom residences that enable to render static content fast. In 2017, it used to be one of the major Google's initiatives of cellular search. We'll see if there's any future for this.

Quick tips:

- Test the responsivity of your internet site in the Mobile-Friendly Test
- Monitor your keyword rankings in mobile search results
- Make positive the cellular model of your website works like a charm

4. Sitemap

A sitemap helps search engines to crawl your content. It's a file where all website sections are listed. It's desirable to have one mainly when you run a large internet site with a difficult shape or when you use wealthy media content.

Having a sitemap doesn't suggest your rankings will improve. According to Google, it's a benefit but you'll never be penalized for not having one.

Quick tips:

- Not all the web sites need a sitemap
- The Sitemap shouldn't include greater than 50,000 URLs and can't exceed 50 MB
- Place the sitemap in the root listing of the website:

https://example.com/sitemap.xml

5. Robots.txt

Robots.txt is a file that tells crawlers which website sections you don't favor to be accessed. It's positioned in https://example.com/robots.txt and it's public.

It's available when you don't prefer some scripts, needless documents or pictures to be indexed.

robots.txt syntax:

User-agent: * (e.g. Googlebot)

Disallow: / (e.g. /images/pizza.png)

Quick tips:

- Don't use robots.txt to cover content material from search engines
- Violating or malware robots are in a position to bypass robots.txt

Further technical search engine optimisation hacks

SEOSiteCheckup is a first-rate tool for on-page SEO analysis. You can analyze one URL per day and download a PDF report barring registering.

Use Google Tag Manager for superior tag management, so you don't want any assistance from internet developers.

HTTPS vs. HTTP: Back in 2014, Google announced they will raise rankings of HTTPS/SSL websites. Today, we recognize that it's a lightweight ranking thing influencing a small percentage of search queries.

However, security is a robust psychological factor. Google Chrome, for example, labels a internet site that is not encrypted with SSL as "not secure", which influences the consumer engagement.

URL/IP canonicalization: IP canonicalization is necessary when a internet site is indexed beneath both its IP address and domain name. URL canonicalization potential that:

```
https://example.com and https://www.example.com/ should resolve to the same URL
```

Chapter 4

Content & SEO

Are content & website positioning two impartial terms or one ideal couple? A lot of entrepreneurs used to suppose that they are separate players. Let's discover out how you can gain from their synergy.

There used to be a famous delusion amongst some entrepreneurs that ideal content material doesn't need SEO. The fact is they have been wrong. And we'll exhibit you why.

Quick tip:

Stop wondering in terms of "SEO vs. content marketing" and begin exploring how well they operate together.

Can you imagine brilliant content material barring on-page optimization or except a single backlink? Likewise, can you imagine a flawlessly optimized internet site packed with content that no one would read?

There used to be regular opinions on search engine optimization and content:

- SEO is for search engines
- Content is for human beings

You should neglect about them! Content & web optimization overlap. You sincerely have to create special content material and optimize it for search engines and human beings at the identical time. This technique is occasionally referred to as "SEO content" or "SEO copywriting".

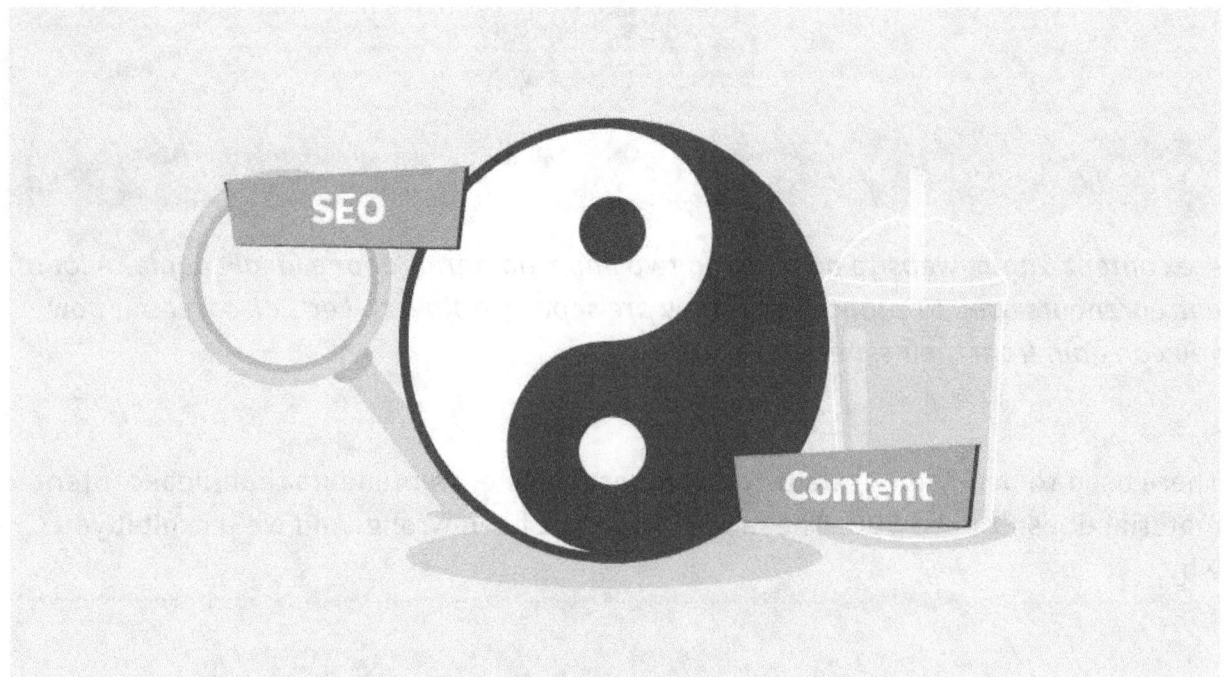

What content material should I create?

This is the first and most vital question. The answer (at least in theory) is simple: "Be unique". Creating original, thought-provoking and attractive content material is a outstanding groundwork for optimization. Content and SEO should go hand in hand from the beginning. You can pick from a range of content types:

Blogs

Blogs are very popular, mainly in ultimate years. A precise weblog is a exquisite source of consumer engagement not solely for bloggers however additionally for e-commerce websites, SaaS groups or professional carrier providers.

Blogging is also a way how to earn cash by way of doing what you love (travel bloggers, advertising bloggers, etc.). Unique blogs with in-depth articles, courses and how-tos can make you a respected influencer in the industry.

Quick tips:

- Stand out from the crowd, be original
- Conduct your personal lookup and use special data

www.creativezib.com

- It's not about length, it's about nice and added value for readers
- Be consistent to create a strong private brand
- Cooperate with businesses and influencers

Product pages

Quality product pages should be the pinnacle precedence of each and every online business. They are regularly used as touchdown pages for PPC campaigns, which include the AdWords remarketing or paid social media.

Quick tips:

- When it comes to eCommerce, simplicity is a need to so genuinely describe the product, how tons is it and how to buy it (CTA)
- Use expert snap shots of products and write appealing copy
- Website pace and UX are more vital than ever

Reviews

Do you desire to be an influencer? Writing a assessment is nothing new however in the world of the internet, the entirety can be reviewed. You can write reviews of tools, films, bars, restaurants, electronics, you title it.

If you manipulate to end up a sincere influencer, you will shape people's opinions and earnings from sponsored reviews. Reviews are an inevitable ingredient of the influencer marketing.

Quick tips:

- Choose one precise topic
- Don't sell your self out simply because some organization can pay well for a 5-star review
- Create your personal fashionable of what is wonderful or poor
- Consistency is the key to become a robust brand

Case studies

Tell the world about how you do things, share fascinating statistics you collected, reveal what is your consumer delight strategy or how you failed when launching a new product. Do experiments, check a variety of myths in your industry.

Probably the pleasant component about case studies is that they are always unique and bring new statistics for readers.

Quick tips:

- Take your time, do your research, put together sufficient unique records and be straightforward (don't lie)
- If you are writing about your business, show the human beings behind the company
- Don't be shy, people aren't involved in dull success stories

Infographics

Everyone likes them! They're easy to share and download. However, authors frequently use them as a text substitution. Infographics then stop up as complex snap shots with a bunch of copy, so they lost the potential.

Creating infographics is about making records fine and effortless to read.

Quick tips:

- Infographics nevertheless have a gorgeous sharing and linking conceivable so make certain to play with each information and design
- Use solely the most vital data
- Create a story, add charts or pictures
- When importing infographics as pictures, don't neglect to write a transcript due to the fact crawlers can't "read" the text in the image

Guides, how-tos, hints & tips

Many times, courses are published as a phase of blogs. The "… guide" and "How to …" are pleasing keywords. You can write publications or how-tos, but preserve in thought that there are already lots of them.

Don't reproduction others, provide different hints & pointers than your competitors.

Quick tips:

- Record your own videos, or at least use customized display screen recordings and screenshots
- Outreach the product proprietor or seller to possibly get a back link or share on their social profiles
- If you are writing a information on your very own product, put the guide on the product landing page to get additional traffic and an interior backlink

Top lists

Another famous structure of written content. There are thousands of pinnacle lists out there so suppose accurately about the topic.

Writing lists is an opportunity to encompass downloadable items to accumulate leads, to earn valuable remarks and to create a buzz thanks to social shares and backlinks.

Quick tips:

- The headline says everything about the post
- Stress gorgeous facts
- Keep the structure simple
- Use validated statistics sources
- Ask the readers to recommend different items to your listing in the comments

Interviews

Getting unique information and opinions from an industry guru is excellent! It helps to construct credibility, traffic, social shares and backlinks.

What's more, interviewees with large target audience generate high traffic for free if they share the interview. You can write, record a video or a podcast.

Quick tips:

- Try to interview a thinking leader (famous individual in the industry)
- Set a clear structure of the interview
- Questions should float naturally
- Allow some space for the interviewee, readers are curious about their thoughts

Videos

Higher engagement, social shares, likes, backlinks, greater leads and conversions. These are the largest advantages of the use of video content. Videos can make bigger conversion quotes by 80%.

Quick tips:

- Write a script, proofread and practice the script
- Prepare some finances both for hardware and enhancing software or for hiring a expert video maker
- Include transcript
- Formulate a advertising and marketing graph or at least design the fundamental promo activities to justify the time and cash you've put in

Ebooks

Ebooks are generally used for lead generation. Maybe extra in the B2B industry but there are nevertheless many B2C web sites that require an e mail tackle to download an ebook.

No one offers you the contact small print simply like that. You need to give readers a right reason. Ebooks typically come in PDF and include a lengthy piece of content.

Quick tips:

- Write it as a perfect book: you need a killer topic, title, preview, credible author, proper proofreading and catchy design
- Motivate users to download: use ebooks with unique data and unique hints that can't be determined on your website or when in contrast to competitors

What's the perfect blog put up length?

A few years ago, long posts stuffed with key phrases ranked on the pinnacle positions easily. Somehow, it received to the point, where it appears like there are lots of almost equal articles with the equal keywords.

Creating content material for the sake of content doesn't work anymore.

If you write an article, you need to keep reader's attention. It starts with the title, first paragraph, content type and most importantly the overall content structure. Use <title>, <h1, h2, h3, …> and other HTML tags correctly.

When it comes to time, posts with a 7-minute reading time are ideal according to Medium. When it comes to SEO, there are many studies. Most of them prove that longer posts rank slightly better.

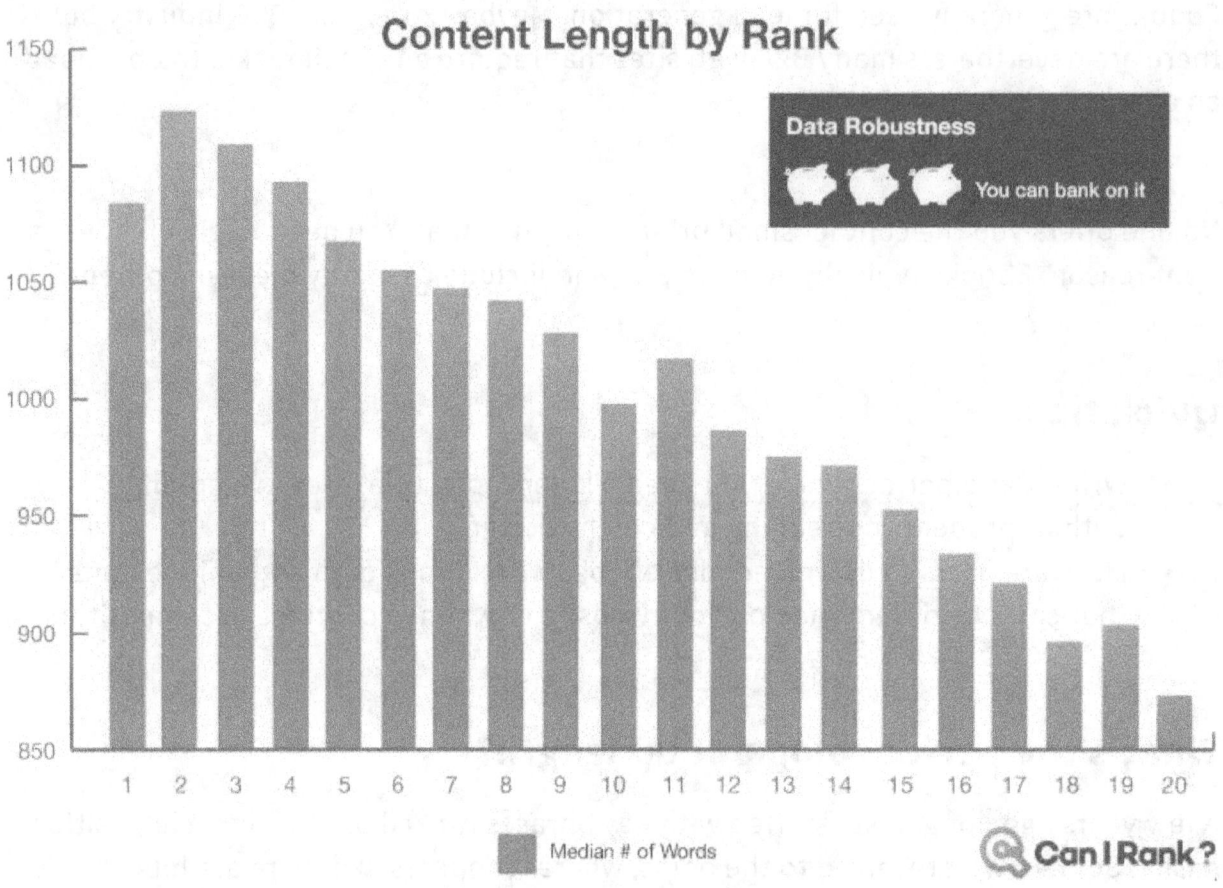

The graph with the correlation between the content length and organic positions, based on the study by Can I Rank.

How about duplicate content in my other articles or website sections?

Sooner or later, we all get to the point the place we need to repeat a few phrases we have written someplace else on our website.

If there's too a whole lot duplicate content across your website, you can use 301 redirects or rel="canonical" link element. You can discover out more on Google Search Console help pages.

Keep calm if you don't graph to copy strains of text. Google algorithm differentiates whether or not you do it on reason or as a herbal phase of your new content.

Just in case you think of copying any individual else's content, you can quit studying this search engine optimisation guide. It's plagiarism and we're strongly against it. The risks, not to point out the costs, of being a copycat are excessive. Google will penalize and skip your internet site from the search results.

Creating brilliant content starts with suitable research. Think of the topic, do key-word research, ask your friends, conduct polls on Facebook or somewhere else.

Once you pick out the topic, scan the market and carefully read all pinnacle ranking websites. Don't replica them, strive to create your own unique content material and optimize it perfectly. Right after that, you can begin promoting it.

Chapter 5

Keyword research

Keyword research honestly belongs to the SEO basics. In this chapter, you will research how to find your area of interest and how to discover profitable key phrases you can rank for.

Creating content material barring keyword lookup doesn't pay off. Your content material will no longer be optimized so the website site visitors will suffer. You definitely have to be aware of what keywords to target to generate adequate first-rate traffic.

Before we start, right here are three elements that shape the key-word research these days:

1. Google Keyword Planner changes

The modifications in Google Keyword Planner (GKP) that passed off in 2016 triggered a bit of an earthquake on the key-word lookup tools market. GKP grew to be worthless for many customers because it stopped displaying actual search volumes for customers who don't spend adequate cash in their AdWords.

In March 2018, Google introduced that they would deactivate AdWords bills with no spending all through remaining 15 months. Approximately at the same time, a new version of GKP was once launched bringing some cool improvements each for search engine optimization and PPC.

2. Keyword optimization over user engagement

In the past, content creators did keyword lookup only to find the keywords with excessive search volumes. They stuffed them into content to trick the search engine algorithms and make sure excessive rankings in organic search. This no longer helps because key-word lookup has grow to be a lot extra complex!

These days you have to work with greater metrics, reflect onconsideration on the Google RankBrain algorithm and the actual SERP you sketch to rank in.

3. Google RankBrain algorithm

Google RankBrain, as hostile to older algorithms, knows 2 things:

How customers engage with the natural search results

What is the search intent (it may also exhibit almost equal results for specific lengthy tail key phrases if the intent is the same)

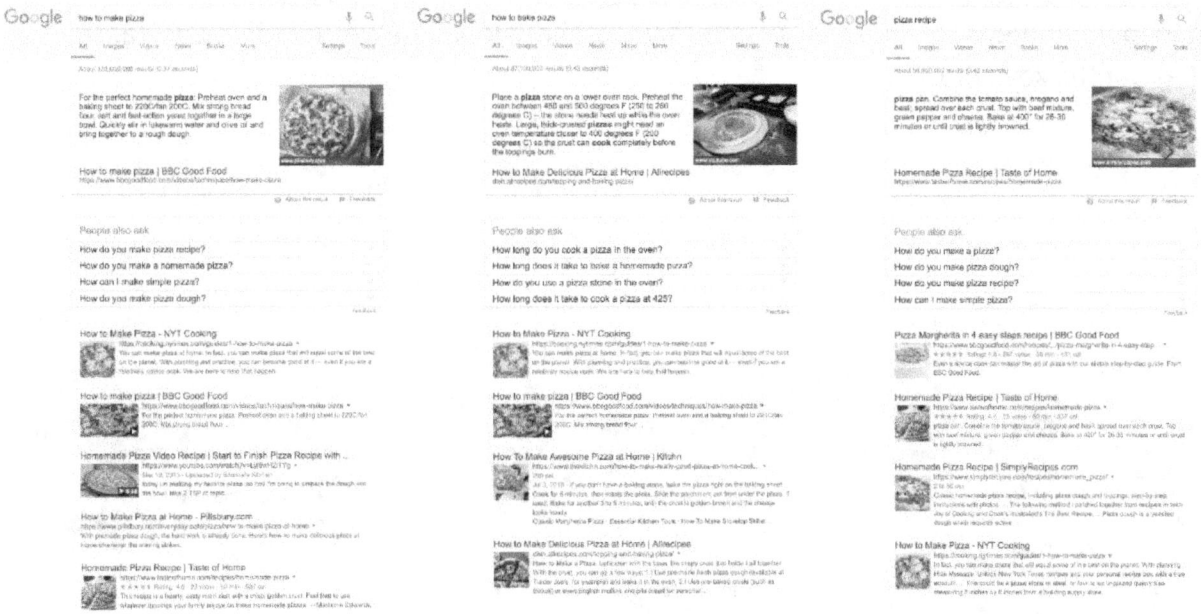

Google RankBrain had been around for a couple of years, however it took some time till it was once wholly implemented. In 2015, Google announced that RankBrain is the third most important ranking signal.

It is a machine getting to know algorithm that varieties the search outcomes through appreciation the search queries and measuring the stage of user engagement.

Search volume, long tail keywords and keyword difficulty

Relevant lengthy tail keywords with excessive search volumes and low keyword challenge – an best mixture of the three most important elements of keyword research. Unfortunately, it is now not usually that effortless and you want to seem for a balance between these factors.

Keep in thinking that you must now not focus solely on how many humans search for the essential keyword. That's because your website will start rating on many associated terms.

Long tail keywords vs. search volumes

Looking for the long tail or middle tail keywords is the place you ought to start. If you run a weblog about pizza recipes, would you like to rank #1 for the key-word "pizza" (with extra than four hundreds of thousands of monthly searches globally)? We wouldn't. Look at the SERP for the keyword.

It wouldn't make sense to be there except you sketch to compete with Pizza Hut or Domino's. Yes, these types of key phrases are very alluring however in most cases, they are now not applicable to your niche. What's more, they characterize solely a few of all the searches throughout the world. The majority of searches are lengthy tail queries.

Long tail keywords have lower search volumes however there are heaps that represent the possibility for you. Count them up and you'll see their large potential.

Visitors who find you by using lengthy tail keywords will have interaction with your content material a lot greater and their conversion charges are higher. It's due to the fact the question is precise ample to find relevant results. And you favor to be at the top of these relevant results.

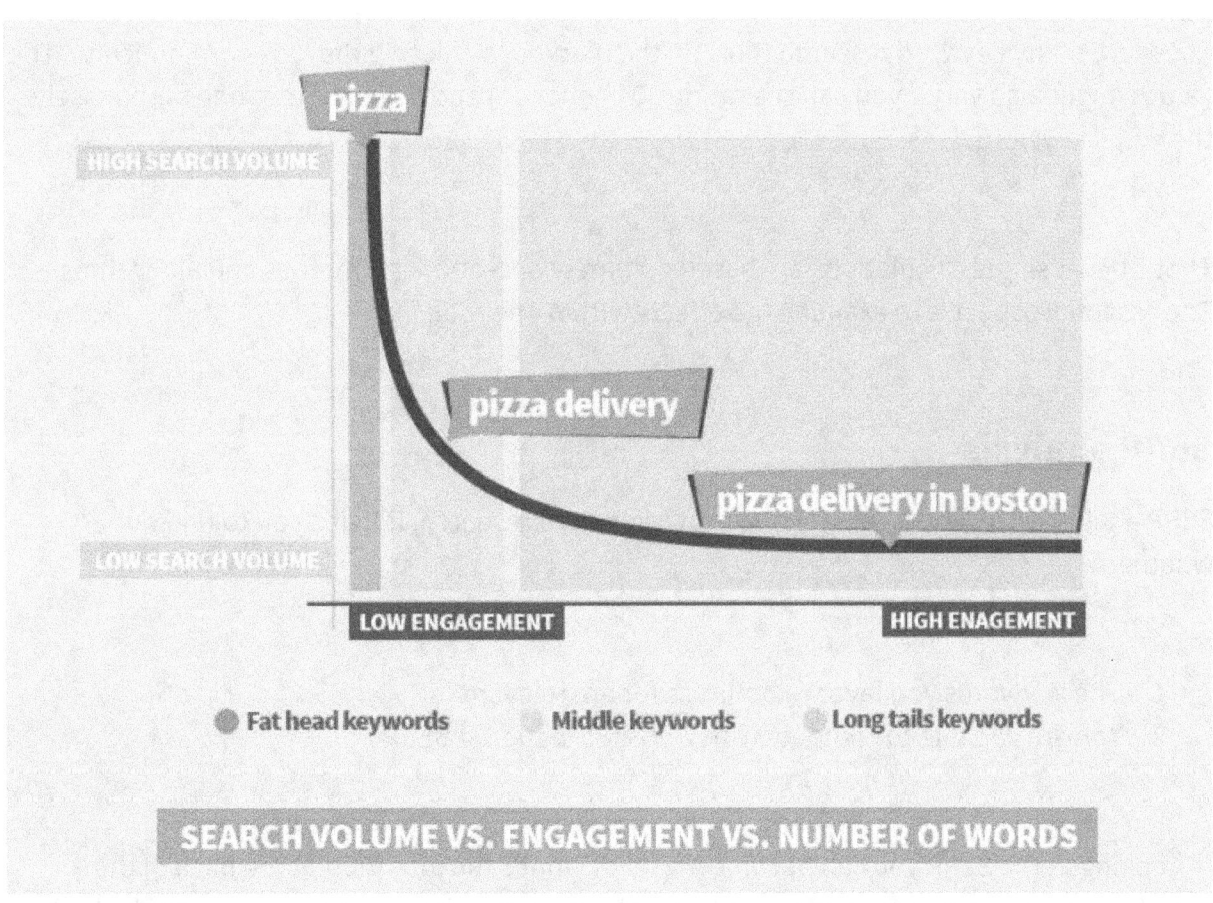

The largest con of long tail key phrases is their search volume. Sometimes, it may also be as low as 100 searches per month. That's why you need to discover the right balance and the metric known as the keyword difficulty will assist you to do reap it.

Keyword difficulty

Once you locate keywords you favor to rank for, you'll need to consider what would it take. The keyword subject or key-word search engine optimisation subject is a very beneficial metric for key-word research.

The cost is normally indicated on a scale from zero to 100 The greater the score is, the harder it is to rank on the 1st SERP for the keyword.

There are many keyword research tools on the market calculating the key-word difficulty. The values might also vary – you can see rating 30 in one tool and 50 in another one for precisely the equal keyword.

That's because the calculations are based on barely one-of-a-kind metrics and algorithms. The essential aspect is to examine the effects within one tool.

SERP analysis

SERP analysis is a very necessary phase of a keyword research. It helps you to find out whether:

1. The keywords you favor to optimize for are relevant
2. You are capable to compete with websites in the 1st SERP

By looking at the SERP you can become aware of what's the intention at the back of the search query. When you are looking out for a "homemade pizza recipe", you probably don't desire to buy a pizza.

Always maintain this in mind so you won't give up up optimizing for incorrect keywords.

Extra recommendations for Google RankBrain keyword research

- Optimize for center and lengthy tail key phrases that aren't longer than 4 or 5 words
- Make certain your website online flawlessly satisfies users' needs. The dwell time (pogo sticking) is now not a ranking factor. It looks like Google doesn't take the time spent on the internet site earlier than bouncing back to the SERP into consideration. It's fair, due to the fact it's now not ample to figure out about the high-quality and relevancy of the search result.
- Create appealing and well-optimized title tags and meta descriptions to expand the CTR from the organic search

How to discover extra niche topics

Besides keyword research tools, you'll discover a lot of subjects and keyword thoughts through the use of the following sources:

- Quora
- Reddit
- industry related forums
- other blogs
- content of competitors
- Google Autocomplete
- feedback from readers

To maintain your lookup organized, write down the topics. Analyze them in the tools and you'll locate many relevant keywords. The extra keywords you have, the more you can analyze and then choose the worthwhile ideas.

Quick tip:

Keep in thought that you need to no longer optimize for dozens of keywords on a single landing page.

You can also check the practicable of a keyword you presently don't rank for in a Google AdWords campaign. Test a variety of key-word matches to get extra thoughts for lengthy tail keywords.

LSI keywords

LSI (Latent Semantic Indexing) keywords are key phrases that are semantically related to the principal seed keyword. They are a amazing add-on to key-word research. They aren't based on any search engine optimisation metric.

LSI key phrases exchange all the time primarily based on the modern-day trend. Adding semantic key phrases to your content is a top idea. You can use LSIGraph to generate a bunch of really useful key-word ideas.

How NOT to do keyword research

Newbies and impatient content creators normally do this:

1. Select a key-word that may be searched.
2. Check its search extent in Google Keyword Planner or any other key-word research device while not questioning about different metrics.
3. Pick one or a few key phrases they in some way think are ok and put them into the heading, sub-headings, each paragraph and meta tags.

The reality is, they're wrong! But don't create content on purpose.

Another mistake is data misinterpretation. A common example is the "Competition" rating in Google Keyword Planner. There's a confusion between this metric and the keyword difficulty.

The opposition in GKP represents the stage of competition in Google AdWords. It doesn't symbolize how difficult it is to rank for that keyword.

Keyword (by relevance)	Avg. monthly searches	Competition
pizza recipe	165,000	Low
pizza dough recipe	246,000	Low
pizza dough	135,000	Low
how to make pizza dough	74,000	Low

Chapter 6
Link building

In the 6th chapter of our SEO for novices guide, we will talk about hyperlink constructing – one of the most vital components of search engine optimization. Let's study how you can build a high-quality link profile.

Link building is a procedure of collecting links from other websites. From the technical factor of view, oneway links are hypertext links that serve as navigation amongst websites. The hyperlinks are crawled by way of search engines.

Why is hyperlink building so important?

Search engines use hyperlinks to discover new websites and to set the usual ranking of a website in SERP. In other words, they explore new content and decide the authority of a internet site based totally on the authority handed from external sources.

This means that a website's hyperlink profile remains one of the most precious ranking factors.

We can distinguish between exterior link building and internal hyperlink building. Internal link constructing means using a link pointing from one web page to every other web page on the identical domain. This web optimization guide explains obtaining hyperlinks from exterior domains.

Types of backlinks

Generally speaking, there are two kinds of backlinks:

- Do observe (authority passing)
- No follow

Do follow backlinks pass the exterior supply authority to the linked website. Imagine the inbound link as a point that boosts search engine marketing of the website, frequently referred to as the "link juice". Do follow hyperlinks appear like this in the HTML code:

```
<a href="http://example.com">Link Text</a>
```

No follow backlinks don't score any points to the linked website. They don't pass the external source authority because of the specific HTML tag that tells crawlers not to count it:

```
<a href="http://example.com" rel="nofollow">Link Text</a>
```

When Google introduced PageRank in 1990, the range of back links was used as an necessary metric of the basic ranking. The extra links you earned, the higher used to be your ranking.

SEO experts began to abuse PageRank in order to amplify the rankings. Many Google algorithm updates led to the fact that many historical hyperlink constructing strategies are now penalized or no longer relevant.

Anchor text

The anchor text is the strongest indicator used via search engines when it comes to backlinks. It's a small piece of textual content hyperlinked to a internet site so it definitely suggests what the website is about.

Therefore, it determines what keywords the internet site have to rank for. Keep in thinking that over-optimized anchors might also lead to an algorithmic penalty by means of Google.

There ought to be a balance amongst the following kinds of anchors:

- Keywords and phrases ("SEO tools")
- Brands ("mangools")
- Branded phrases ("SEO equipment by mangools")
- Generic anchors ("page")
- Naked URLs ("mangools.com)
- CTAs ("click here, "read more")

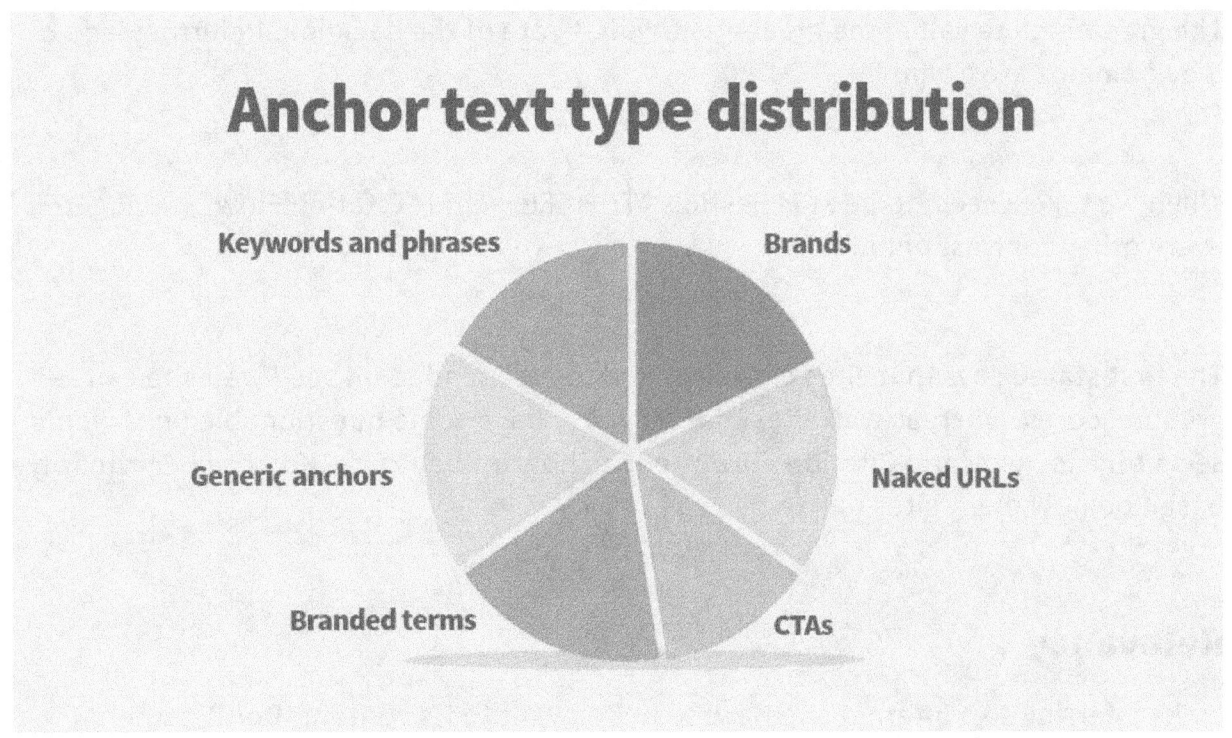

For more precise statistics about anchor texts and how to optimize them, check our SEOpedia's article dedicated to anchor texts.

Quality components of backlinks

Google considers a couple of first-rate aspects of the referring websites:

- Domain Authority
- Page Authority
- Other link profile excellent metrics

There are various metrics by means of Moz and Majestic which help us to evaluate/approximate these qualities:

- Moz Page authority (PA)
- Moz Domain authority (DA)
- Majestic Trust Flow (TF)
- Majestic Citation Flow (CF)

The greater is the value, the greater is the first-rate of the backlink. Referring .edu or .gov domains are a win!

When we take a seem to be at Majestic's "Trust Flow and "Citation Flow", we'll get every other contrast of links.

The first stated says that sites closely linked to a depended on seed web site can see greater scores, whereas web sites that may also have some questionable links would see a tons lower score. Citation Flow predicts how influential a URL may be primarily based on how many sites link to it.

Relevancy

Links referring to a internet site have to be relevant to its content. Don't gather a backlink from a clothes e-shop if you write a weblog about pizza.

Link placement

Links positioned in the foremost articles or sections are higher than links in footers and sidebars. Single links tend to be extra valuable than sitewide links.

Sitewide links appear on all pages of a website. They are usually in the footer, header, sidebar or blogrolls. Sitewide links are super each for internal and exterior link building.

They can generate a lot of traffic. Don't get scared by means of them however make sure to use solely natural links and hold in thought that their website positioning workable might also be a bit decrease when in contrast to single links.

Besides the factors referred to above, we need to reflect onconsideration on the freshness of the link, the anchor quality, recognition and social signals.

Link building techniques

It's not effortless to accumulate a superb backlink. The techniques that were the easiest (reciprocal hyperlinks or directory submissions) do now not work anymore, so the SEOs spend a lot of time trying extraordinary approaches.

These are the most popular link building ideas:

- Content-based hyperlink building: Create content material that will naturally appeal to backlinks, social shares and referral traffic
- Get social as a great deal as possible: Share your content on social media, promote it on Facebook, join discussions, comment applicable posts and create connections
- Guest posts are productive sources of relevant backlinks
- PR articles written with the aid of gurus and posted on information portals will supply you extremely good back links but prepare some finances for this and make positive they are in reality relevant
- Buying oneway links by way of paid weblog posts
- Backlinks from forums, Q&A sites, top lists, comments, content aggregators, business listings, etc. Keep in idea that substantial majority are no comply with or low-quality backlinks
- Find out what works for others: Spy on your opponents to explore their backlinks by using equipment such as LinkMiner or others
- Do email outreach: A time-consuming but still quite high-quality technique. Check the websites that link to your competitors, create better content and contact relevant human beings at the back of these web sites to hyperlink to your website instead
- Grey/black hat techniques such as PBN (Private Blog Network)

All in all, the best links are natural links. On the other hand, to acquire them may not be easy at all.

Grey hat, black hat techniques and penalties

Paid inbound links and PBN (Private Blog Network) are any other way to construct backlinks however you have to be very careful. Make sure they appear as herbal as possible.

Private blog networks are websites to put into effect links to other web sites (ones that generate money). It's an luxurious technique. You want to pay for a domain, internet hosting and set of weblog posts.

Moreover, it's volatile because Google may additionally ban them if you are not cautious enough. Sometimes the charges of a PBN can be higher than the profit.

Penalties

Google Penguin algorithm replace from April 2012 introduced penalization for bad, spammy, or low-quality links.

However, if you have spammy or low-quality backlinks, don't panic! Google shouldn't consider them a bad ranking factor. But simply in case, you can nevertheless disavow such one way links in the Search Console.

Best practices

- Do link constructing regularly, it's not a one-time effort
- Acquire a few first-rate links alternatively than many low-quality or spammy links
- Do natural anchor text distribution based totally on the guidelines we listed in the "Anchor text" section
- Avoid one-way links from spammy web sites and networks
- Try to accumulate back-link placements which may want to carry you referral site visitors as well
- Focus on competitors' back links you can replicate without problems

Chapter 7

UX & SEO

UX and search engine marketing are a very dynamic duo. In this chapter, we focus on their synergy so you will discover out how you can positively have an effect on your organic rankings by doing a little more optimization.

The analytical part of the search engine optimization is about monitoring natural key-word positions, internet site site visitors and its engagement in Google Analytics, Webmaster Tools and many other specialised tools.

UX (user experience), on the other hand, basically deals with design, wireframes, projections and techniques for flawless patron experience. If you favor to advantage the most, you have to look for their synergy.

While SEOs want to recognize it's not solely about rankings, UX experts have to admit that user trip kicks in even before the usage of the website.

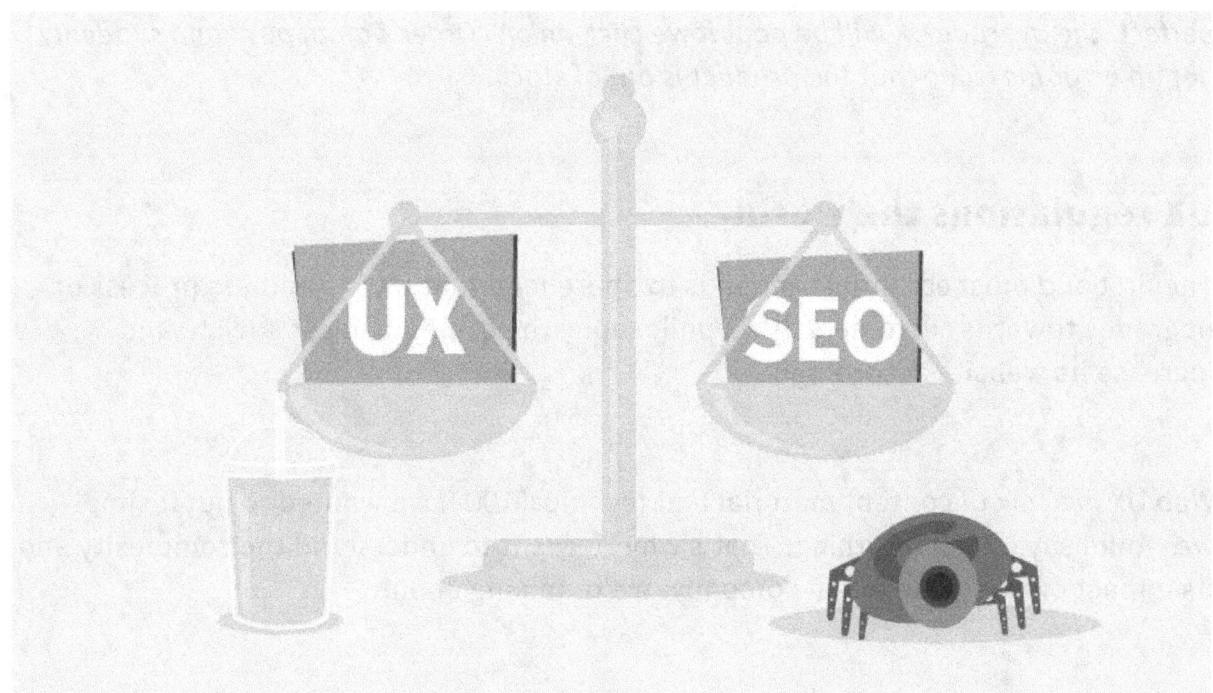

User experience. What is it?

User experience is basically every user's interaction with the company, its internet site or products. It's the ordinary trip influencing product development, design, advertising and customer support.

In this search engine optimisation guide, we focus on the "online" consumer experience and its relationship with SEO. UX is no longer only about assembly customer requirements. It's about going beyond these requirements.

You want to assume out of the container and improve a internet site that naturally covers every viable want of a customer barring   bothering him.

Quick tip:

UX is no longer UI (user interface). Yes, it's a phase of it but we want to distinguish these terms. Imagine an on line store. Despite its UI for finding a product can be perfect, the average UX will be poor if we click on on "Order & pay" and suddenly get an error message that the product is out of stock.

UX regulations the world

The first and most essential section is to apprehend that UX is a endless process of upgrades towards customers. It's a philosophy the company must stick to and increase its website accordingly.

Web UX is a mix of content material and technical UX. If we wanted to put it simply, we would say UX is everything. That's why we want to understand the complexity and its impact on everyone in the company. We're talking about:

- Website structure
- Website navigation
- Design
- Web development
- Content excellent and density
- Content structure
- Technical search engine marketing aspects
- Real-time client support
- Smooth workflow
- Ability to attain the conversion stage easily

UX and SEO challenges

Even if we think "UX and SEO" as a substitute of "UX vs. SEO", we have to cope with challenges every as soon as in a while. Here are some of the most frequent clashes.

Website

Let's put the whole lot on a single web page to make certain smooth UX. This is a legitimate thought in the UX world. It's convenient and it makes sense. But SEOs know that organic visitors is at least the 1/2 of the success.

Many site visitors should come from the organic search or get admission to the page immediately primarily based on their preceding actions. Other visitors are referred from a variety of sources such as social media, blogs or PPC campaigns.

Can you imagine a single page journey for all of these unique intentions and search queries? People use Google to discover what they want. Your internet site have to be the answer.

Website structure

Based on the preceding point, SEOs will assume about more than a few landing pages with applicable keywords, meta tags and the standard internet site shape from the website positioning point of view. The UX specialist will are seeking for a minimal onsite navigation.

The truth is that on-site navigation hacks such as search bar with wise autocomplete, interior linking with anchor texts or immediately purchaser help improve each UX and SEO.

Infographics

This is a very frequent UX vs. website positioning case. Infographics are a excellent way to make information seem nice. However, from the search engine optimization point of view, infographics uploaded as snap shots have a very low value. In this case, don't overlook to include the content material in the website's copy.

Videos

A video is a in all likelihood the quickest and most engageable way how to share content. Moreover, it helps website positioning via gaining backlinks, likes, remarks or shares. But it can eat up all treasured keywords. That's why we ought to include a transcript on the website.

How to measure UX and SEO

From Google Analytics thru specialized reports to interior data, there's a variety of choices we can use.

Engagement metrics and conduct in Google Analytics

"Bounce rate", "Pages per session" and "Avg. session duration" furnish the simple view on user interaction. These metrics are nearly in every file in GA, so we can see them for a number of traffic sources.

Behavior		
Bounce Rate	Pages / Session	Avg. Session Duration
26.98% Avg for View: 26.98% (0.00%)	4.88 Avg for View: 4.88 (0.00%)	00:07:18 Avg for View: 00:07:18 (0.00%)
29.13%	4.84	00:07:14
22.49%	4.95	00:07:26

Engagement metrics in Google Analytics "Behavior"

The greater specific data is accessible in the category "Behavior". Here we can see how visitors engage with various sections of the website. It's up to us whether we prefer a record based on site visitors sources, landing pages or something else.

"Behavior flow" comes with available advice. It tells us how many customers went via or dropped-off when traveling that unique section.

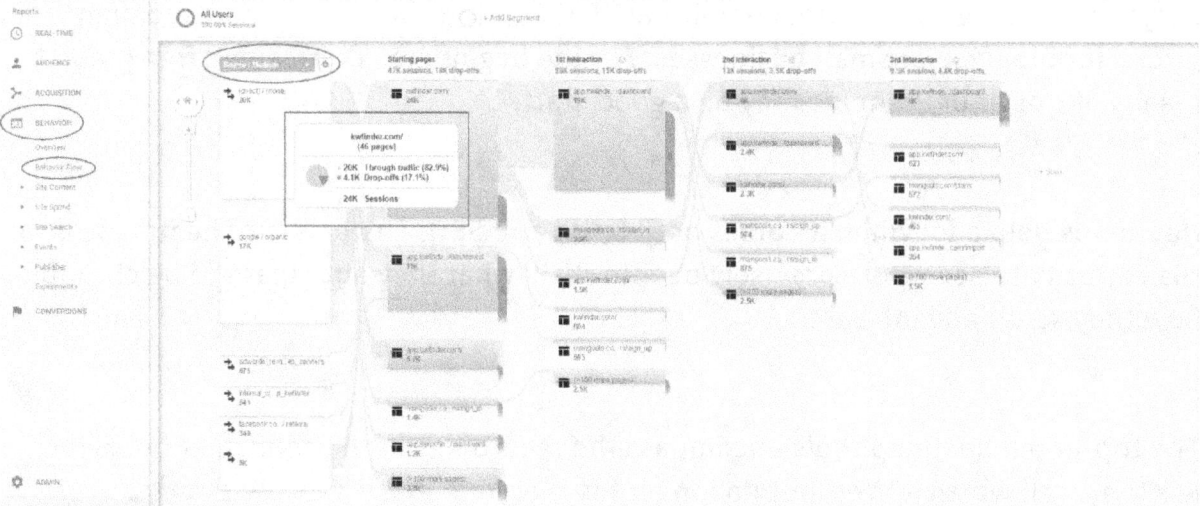

Conversion funnels in Google Analytics

This is very useful if a conversion method consists of more than one step. It can be a registration form, buying method or subscription form. This document will inform us the specific wide variety of traffic who went via the process. It's just another extremely beneficial information for UX optimization.

Conversion funnel report in Google Analytics

Heatmaps & recordings

Heatmaps essentially visualize all the records we see in Google Analytics whilst presenting many beneficial reports. Sophisticated heatmap equipment such as Hotjar or Crazyegg come with complex analytic options but they can be too high priced for bloggers or small businesses. For the beginners, a restrained free graph by means of Hotjar or Ptengine may be a good start.

Heatmaps help a lot when it comes both to UX and SEO. The power of heatmaps is in the capacity to see where precisely people click, what they anticipate to be clickable, how they scroll and interact.

The top-notch heatmap tools encompass the recordings of the consumer behavior. Are they well worth it? Yes, but no longer for everyone. Businesses with extra complicated (e-commerce) pages (and normally with greater budgets) need to clearly go for it, however there is no factor for bloggers to spend hundreds of bucks on these expert tools.

Internal data

Do you be aware of what's high-quality about interior data? There aren't any additional costs. Your colleagues or you create reports packed with useful records all the time:

- Customer questions
- Feedback forms
- Comments and reactions on social media
- Customer desires and complaints
- Marketing reports
- Email responses
- Reports from web developers

Testing and site speed

Testing the website before its launch is a must! Beta testers grant a lot of useful information. They're the first real users. On pinnacle of that, you can do some usability checking out with colleagues, pals or followers during the improvement process.

Pay attention to key factors such as site speed, mobile optimization, and different vital technical search engine marketing aspects. The complete listing with a downloadable infographic can be found in the third chapter.

Chapter 8
SEO resources

Besides this search engine optimization information for beginners, there are many super sources that will help you to study web optimization and enhance your optimization skills. We have picked the top ones for you!

SEO is a developing industry and there are actually thousands of sources you can study from – wheter you choose blogs, ebooks, infographics, boards or following the experts. The general on-line advertising organization Ignite Visibility even made an website positioning Movie!

Of course, too many resources can be a little overwhelming for an search engine optimisation novice as it may additionally be difficult to distinguish exceptional content material from the awful one. That is why we determined to dedicate the ultimate chapter of our website positioning information for beginners to the nice search engine marketing resouces we hand-picked personally!

www.ingramcontent.com/pod-product-compliance
Lightning Source LLC
Chambersburg PA
CBHW080825170526
45158CB00009B/2532